Where Did the Water Go?

by Amy S. Hansen

Science Content Editor:
Kristi Lew

rourkeeducationalmedia.com

Scan for Related Titles
and Teacher Resources

Science content editor: Kristi Lew
A former high school teacher with a background in biochemistry and more than 10 years of experience in cytogenetic laboratories, Kristi Lew specializes in taking complex scientific information and making it fun and interesting for scientists and non-scientists alike. She is the author of more than 20 science books for children and teachers.

© 2012 Rourke Educational Media

All rights reserved. No part of this book may be reproduced or utilized in any form or by any means, electronic or mechanical including photocopying, recording, or by any information storage and retrieval system without permission in writing from the publisher.

www.rourkeeducationalmedia.com

Photo credits: Cover © CAN BALCIOGLU; Cover logo frog © Eric Pohl, test tube © Sergey Lazarev; Table Of Contents © Burnedflowers; Page 5 © Ben Heys; Page 7 © Ruslan Nabiyev; Page 9 © CAN BALCIOGLU; Page 11 © Oskar Orsag; Page 12/13 © Burnedflowers; Page 14 © Monika Hunácková, Christian Lopetz; Page 15 © Peter Hulla Page 16 © Monika Hunácková,Christian Lopetz; Page 17 © Aaron Amat; Page 18 © Christian Lopetz; Page 19 © Fedor Kondratenko; Page 20 © dinadesign; Page 21 © Monika Hunácková;

Editor: Kelli Hicks

Cover and page design by Nicola Stratford, bdpublishing.com

Library of Congress Cataloging-in-Publication Data

Hansen, Amy.
 My science library / Amy S. Hansen.
     p. cm. -- (Where did the water go?)
 Includes bibliographical references and index.
 ISBN 978-1-61741-751-1 (Hard cover) (alk. paper)
 ISBN 978-1-61741-953-9 (Soft cover)
 1. Water--Juvenile literature. 2. Water-supply--Juvenile literature. I. Title.
 GB662.3.H37 2012
 553.7--dc22
                          2011004763

Rourke Educational Media
Printed in the United States of America,
North Mankato, Minnesota

rourkeeducationalmedia.com

customerservice@rourkeeducationalmedia.com • PO Box 643328 Vero Beach, Florida 32964

Table of Contents

Three Forms of Water   4
Why Does Water Change?   8
How Do Water Molecules Change?   12
Show What You Know   22
Glossary   23
Index   24

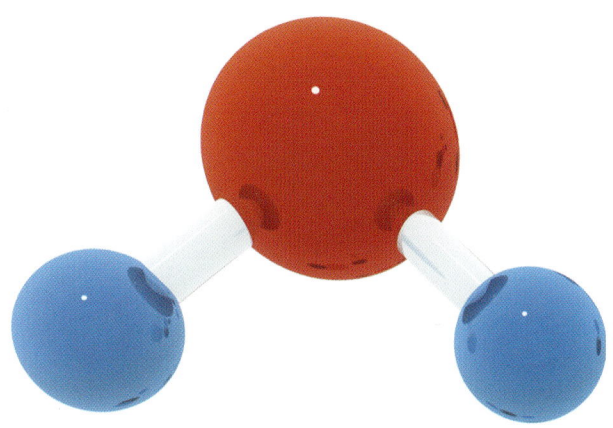

Three Forms of Water

Have you ever spilled ice and not cleaned it up right away? What happens to the ice? The ice **melts** and you're left with a puddle of water.

Ice melts as it warms up.

If you don't clean up the puddle, what happens? You guessed it. The puddle dries up. No more spill!

You've just seen the three forms of water. First it was a **solid**. Then it melted to the **liquid** form of water. And finally, it became a **gas** called **water vapor**.

Have you ever wondered why you have to keep adding water to a fish tank or swimming pool? Where is the water going? The answer is it's in the air and you're seeing evaporation at work.

Why Does Water Change?

Water changes its form when the **temperature** changes. When water is very cold, water is a solid. When it is warm, water is a liquid. When it is hot, water boils and part of it becomes a gas.

This kind of water vapor is very hot. It spreads out quickly.

Water can also become a gas when it isn't hot or boiling. If the air is dry, water will become a gas at lower temperatures. This is what happened to the water in your puddle.

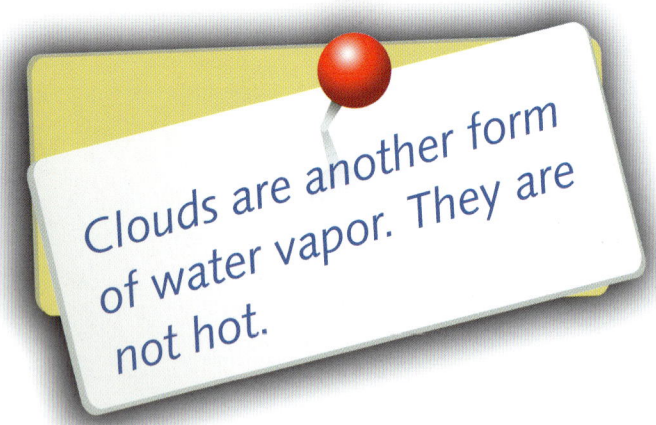
Clouds are another form of water vapor. They are not hot.

How Do Water Molecules Change?

Water is made of tiny units called **molecules**. Molecules are so small you would need a super-strong microscope to see them. The molecules are the same in each form of water, but they are arranged differently.

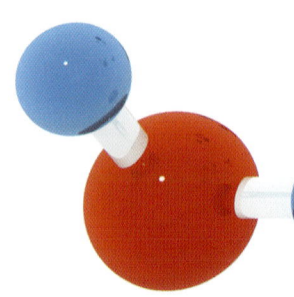

water molecules

Do you know why water is called H$_2$O? Every water molecule is made of two hydrogen atoms and one oxygen atom. The abbreviation for this is H$_2$O.

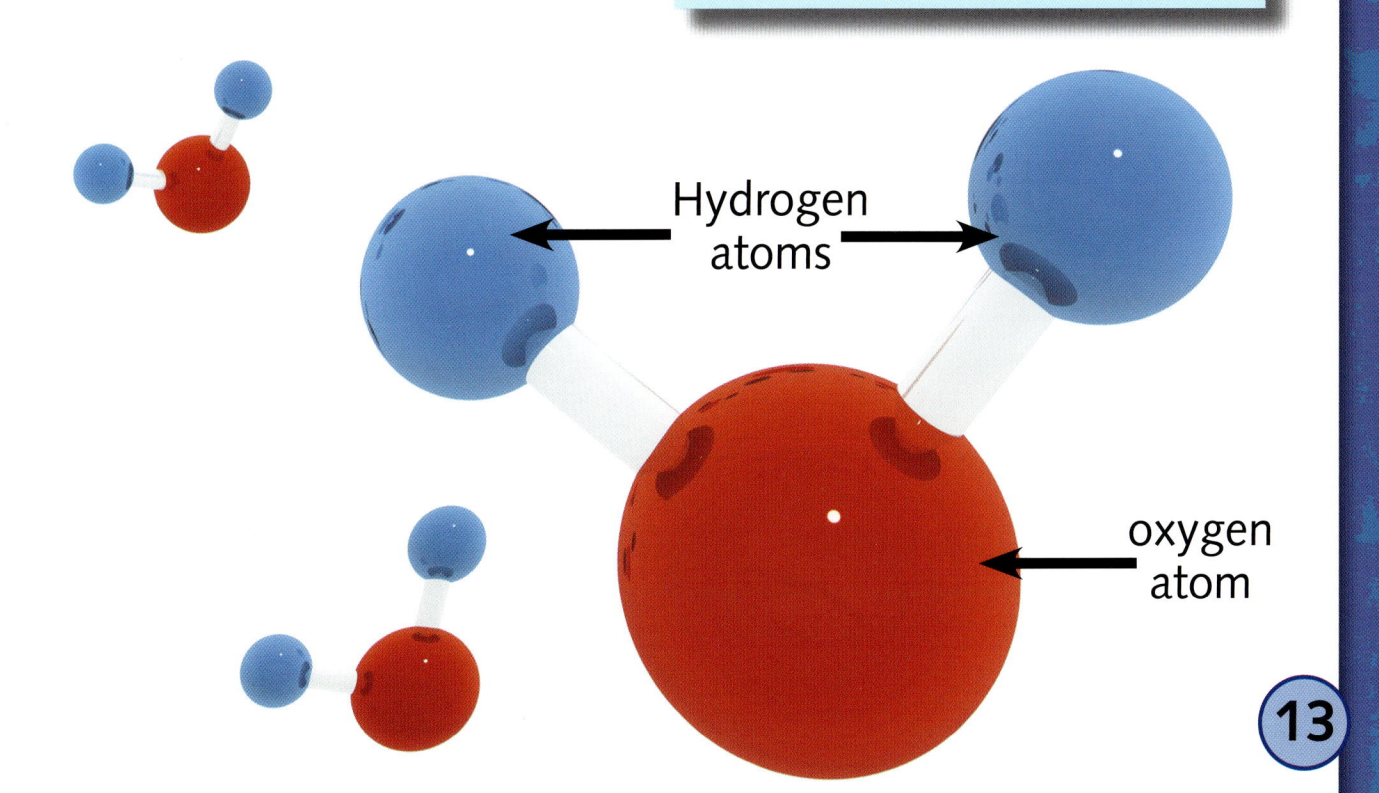

← Hydrogen atoms →

← oxygen atom

13

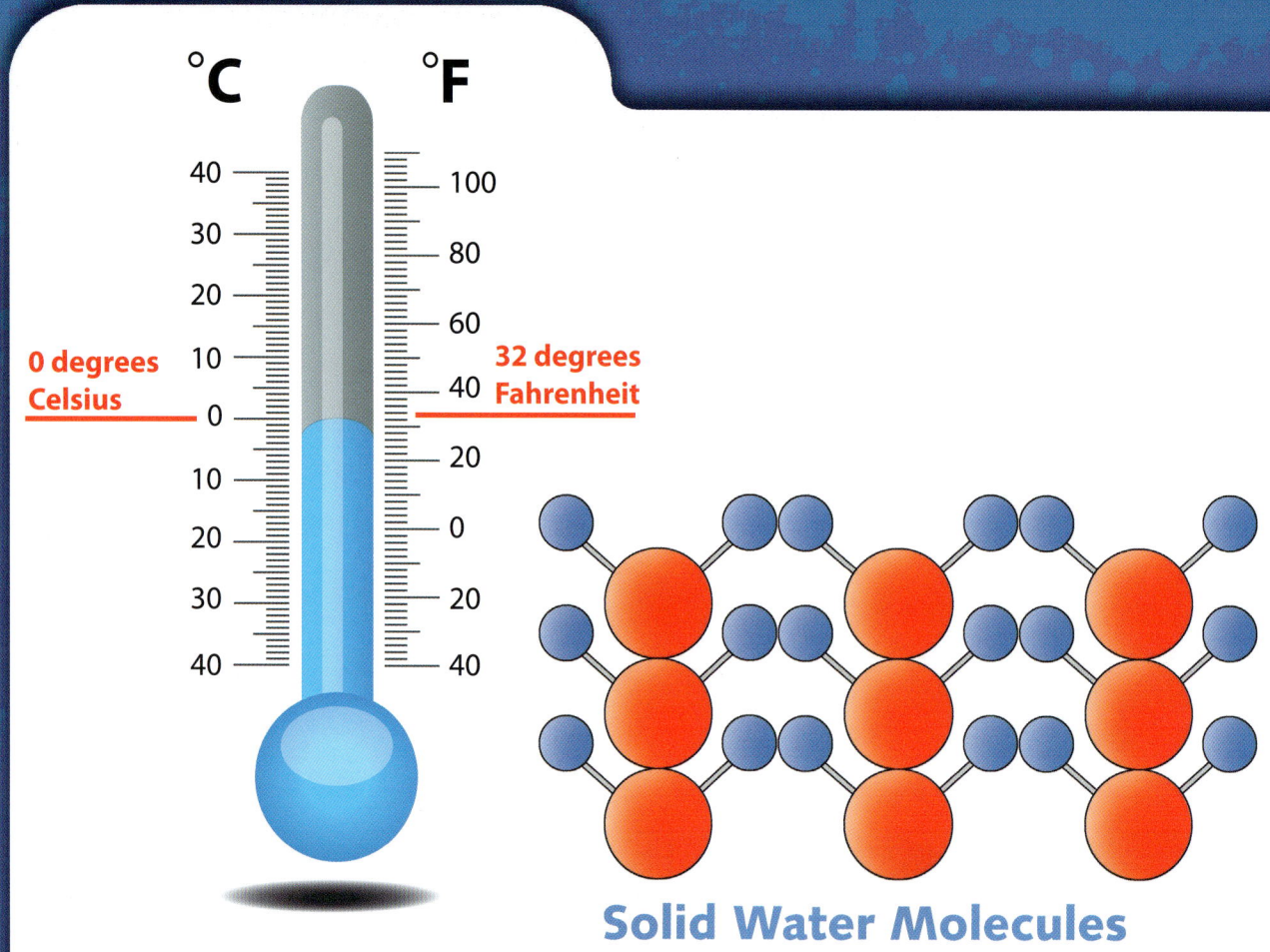

Solid Water Molecules

When water is cold, it is a solid called ice. The water molecules line up. They are cold so they hardly move. The solid holds its shape.

These icicles will hold their shape until they melt.

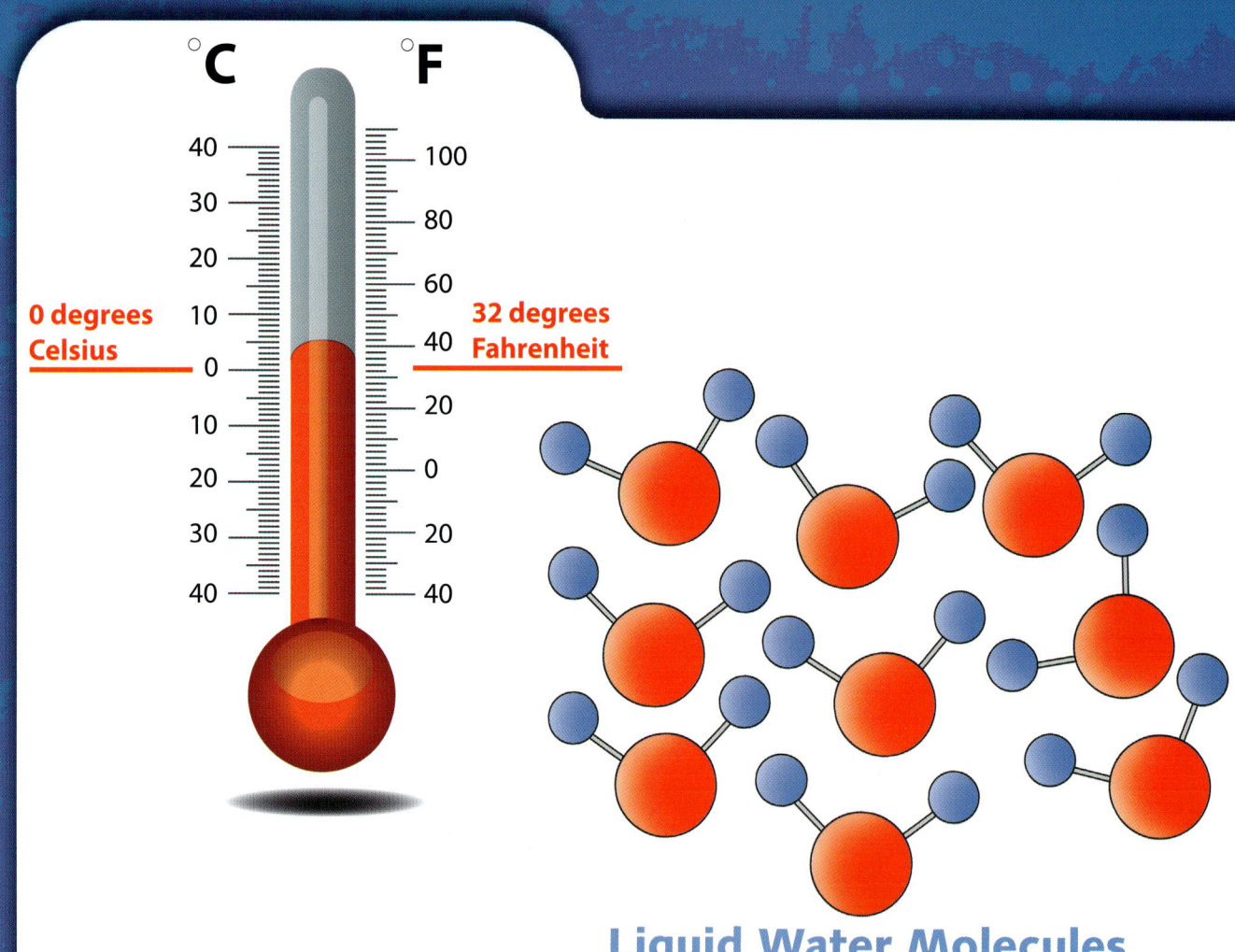

Liquid Water Molecules

When water is warm, it is a liquid. The water molecules do not line up. They have some energy from heat, so they move around. A liquid cannot hold its shape.

When water is a liquid, it needs a container to hold its shape.

When water is hot it is a gas called water vapor. The molecules have a lot of energy from the heat. They zip around. The gas takes up all the space it can.

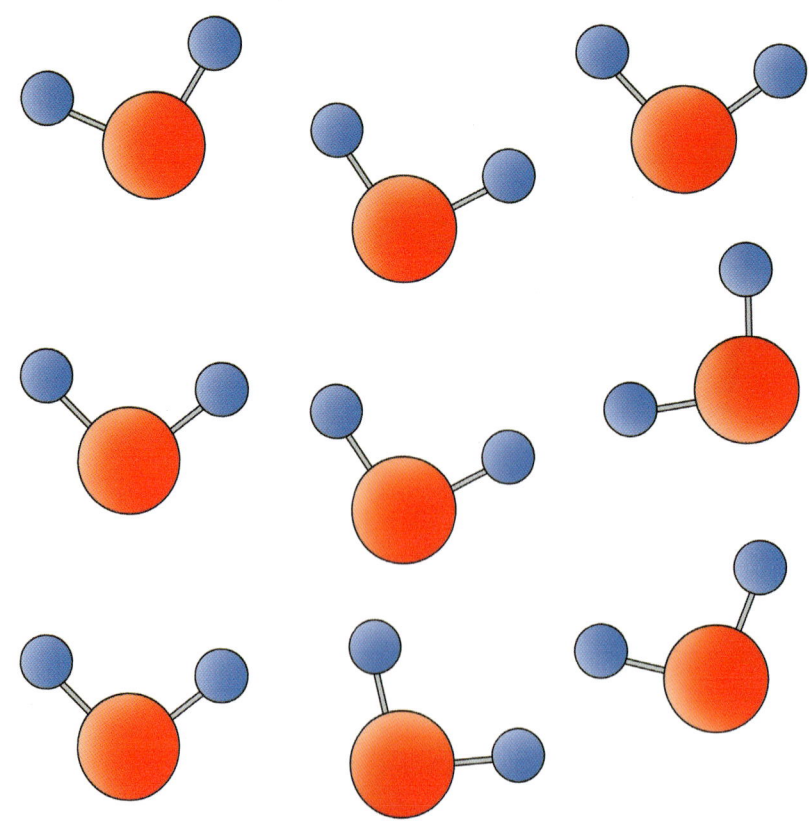

Gas or Water Vapor Molecules

When water is a gas, the molecules move farther apart. Gas expands to fill the space.

Water also becomes a gas if the air is dry. This is why puddles dry out, and why your towel will dry if you spread it out.

What happens when the gas cools down? The molecules slow down. The gas **condenses**. It becomes liquid water.

What happens if we put liquid water in the freezer? The water becomes cold. It **freezes**. The molecules stop moving around. The liquid water becomes ice. Now you know where the water goes.

Which Type of Water?

Solid: Ice
Colder than 32 degrees Fahrenheit (0 degrees Celsius)

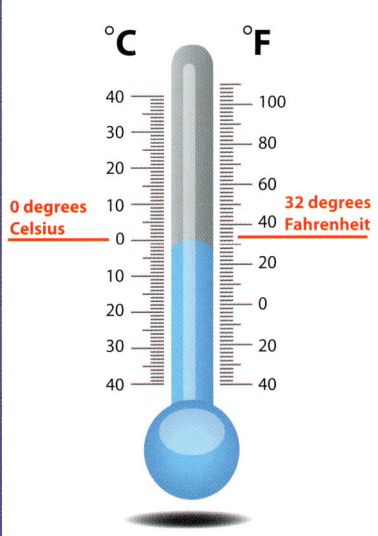

Liquid: Water
Warmer than 32 degrees Fahrenheit (0 degrees Celsius)

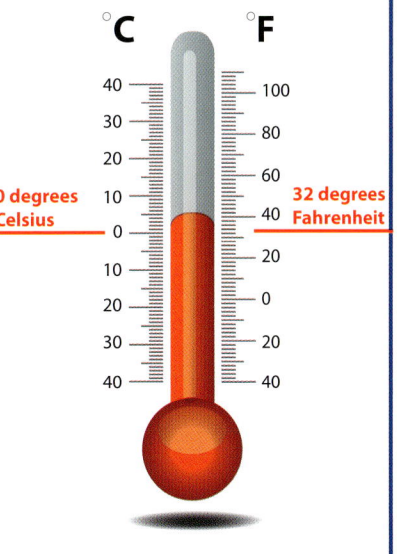

Gas: Water Vapor
Hotter than 212 degrees Fahrenheit (100 degrees Celsius)

or

When the air is dry, liquid water becomes gas at a lower temperature.

Show What You Know

1. Can you think of something that melts other than ice?

2. Where do the gas molecules get their energy to zip around?

3. What makes the molecules slow down?

Glossary

condenses (kuhn-DENS-ez): when gas changes to a liquid, usually through cooling

freezes (FREEZ-ez): changes from a liquid into a solid

gas (GAS): a substance that spreads out to fill the space around it and is often invisible

liquid (LIK-wid): a substance that pours easily

melts (MELTZ): to change from a solid to a liquid

molecules (MAH-luh-kyools): two or more atoms chemically bonded together

solid (SAH-lid): an object that is firm that is not a liquid or a gas

temperature (TEM-pur-uh-chur): the measurement of how hot or cold something is, usually measured with a thermometer

water vapor (WAW-tur VA-pur): a gas formed as liquid

Index

energy   16, 18
gas   18, 19, 20, 21
liquid   6, 8, 16, 17, 20, 21
molecule(s)   12, 13, 14, 16, 18, 19, 20

solid   8, 14, 21
temperature(s)   8, 10, 21
water vapor   6, 9, 10, 18, 21

Websites

www.kids-science-experiments.com/steamingup.html

ga.water.usgs.gov/edu/watercyclecondensation.html

www.kidzone.ws/water/

www.pbs.org/parents/catinthehat/activity_exploring_weather.html

kids.earth.nasa.gov/droplet.html

About the Author

Amy S. Hansen is a science writer who lives in a suburb of Washington, D.C. where the summer air is often filled with so much water vapor that it is muggy and difficult to move.